REPORT

Balancing Rated Personnel Requirements and Inventories

James H. Bigelow, Albert A. Robbert

Prepared for the United States Air Force

The research described in this report was sponsored by the United States Air Force under Contract FA7014-06-C-0001. Further information may be obtained from the Strategic Planning Division, Directorate of Plans, Hq USAF.

Library of Congress Cataloging-in-Publication Data

Bigelow, J. H.
Balancing rated personnel requirements and inventories / James H. Bigelow, Albert A. Robbert.
p. cm.
Includes bibliographical references.
ISBN 978-0-8330-5094-6 (pbk. : alk. paper)
1. United States. Air Force—Officers—Rating of. 2. United States. Air Force—Officers—Supply and demand. 3. Air pilots, Military—Rating of—United States. 4. Air pilots, Military—Supply and demand—United States. 5. United States. Air Force—Personnel management. I. Robbert, Albert A., 1944- II. Title.

UG793.B54 2011
358.4'10332—dc22

2011012225

Published 2011 by the RAND Corporation
1776 Main Street, P.O. Box 2138, Santa Monica, CA 90407-2138
1200 South Hayes Street, Arlington, VA 22202-5050
4570 Fifth Avenue, Suite 600, Pittsburgh, PA 15213-2665
RAND URL: http://www.rand.org/
To order RAND documents or to obtain additional information, contact
Distribution Services: Telephone: (310) 451-7002;
Fax: (310) 451-6915; Email: order@rand.org

Preface

This report documents RAND Corporation research on policies to bring requirements for rated personnel into balance with inventories and subsequently to maintain them in balance. For at least a decade, the Air Force has attempted to achieve this balance by producing and absorbing rated personnel as fast as possible. However, shortages have persisted and have even grown, so the Air Force has shifted emphasis to reducing the requirements for rated personnel.

The project that produced this report is one of a series of Rated Force Management studies cosponsored by the Air Force Deputy Chief of Staff for Air, Space, and Information Operations, Plans and Requirements (AF/A3/5), and the Deputy Chief of Staff for Manpower and Personnel (AF/A1). The study was conducted within the Manpower, Personnel, and Training Program of RAND Project AIR FORCE.

Related work is reported in the following publications:

- *Fighter Drawdown Dynamics: Effects on Aircrew Inventories*, William W. Taylor, James H. Bigelow, and John A. Ausink (MG-855-AF).
- *Absorbing Air Force Fighter Pilots: Parameters, Problems, and Policy Options*, William W. Taylor, James H. Bigelow, S. Craig Moore, Leslie Wickman, Brent Thomas, and Richard S. Marken (MR-1550-AF).

This report is intended to assist senior Air Force policymakers in developing policies that will maintain a balance between requirements for and inventory of rated personnel. The reader is assumed to be familiar with the Air Force and, in particular, with aircrew management.

RAND Project AIR FORCE

RAND Project AIR FORCE (PAF), a division of the RAND Corporation, is the U.S. Air Force's federally funded research and development center for studies and analyses. PAF provides the Air Force with independent analyses of policy alternatives affecting the development, employment, combat readiness, and support of current and future aerospace forces. Research is conducted in four programs: Force Modernization and Employment; Manpower, Personnel, and Training; Resource Management; and Strategy and Doctrine.

Additional information about PAF is available on our website:
http://www.rand.org/paf/

Contents

Figures and Tables

Figures

Tables

Summary

For more than a decade, the Air Force has experienced shortages of rated officers. Since the early 1990s, force structure has declined over 50 percent, reducing the capacity to produce and absorb new rated officers. Requirements for rated officers have declined as well, but the Air Force has not been able to reduce nonflying rated billets (most of which are staff positions) in proportion to the force structure reductions. As a consequence, the Air Force has attempted to produce and absorb rated officers at the maximum possible rate.

The effort has not been enough. At times, the overall inventory of rated officers has been sufficient to fill overall requirements, but there have always been specific categories—fighter pilots, in particular—in which large shortages have been a way of life. Even the overall picture has deteriorated in the past year or two, as new requirements have emerged for categories such as unmanned aircraft systems (UASs), new special operations forces aircraft, and the creation of Air Force Global Strike Command.

In February 2009, the Vice Chief of Staff of the Air Force chartered the Rated Staff Requirements Integrated Process Team (IPT) to recommend courses of action for (1) balancing rated staff requirements with rated inventory and (2) subsequently maintaining them in balance. Because the inventory has been made as large as possible, the IPT had to reduce the number of positions to which rated officers are assigned. However, because rated staff positions have been reviewed repeatedly and found to be valid requirements, the IPT rejected the notion of eliminating requirements. Instead, it directed the owners of rated positions—major commands, field operating agencies, direct reporting units, joint agencies, Headquarters Air Force, and the Secretary of the Air Force—to recategorize specified numbers of staff positions. The owners, in other words, were instructed to find people other than active rated officers to fill those positions. The replacements could be civil servants or contractors (particularly individuals with prior rated experience in uniform), members of the Air National Guard or Air Force Reserve, active nonrated officers, or enlisted personnel.

Owners were able to recategorize 836 positions, enough that requirements and inventory projected for the end of fiscal year (FY) 2010 are nearly in balance, and progress is being made in filling the recategorized positions.

The remaining task, and the primary focus of this report, is to devise a process that will maintain the balance between rated requirements and inventory over the long term. This process should include the following five actions:

1. The Air Force should institutionalize a version of the recategorization process pioneered by the Rated Staff Requirements IPT. Owners currently conduct an annual review of all rated positions to ensure that they require rated expertise and are necessary for accom-

plishing the Air Force mission. But the review takes no notice of possible inventory shortages and should therefore be changed to account for such shortages. Each owner should be given a rated authorization quota for each category of rated officers (i.e., separate quotas for fighter pilots, bomber pilots, etc.) and should be prohibited from labeling a position "authorized" unless it falls within the quota. Owners could trade quotas among themselves and could recategorize positions that do not fall within their quotas. (See pp. 9–12.)

2. The Air Force must streamline the processes for converting the recategorized positions. The IPT arranged to include funding in the current program objective memorandum for 572 civilian positions by the end of FY 2013. Some analysis has also been conducted to identify Air Force specialty codes that could have some of their members assigned to formerly rated staff positions. But work remains. (See pp. 13–15.)
3. The Air Force should plan for the effects of major actions on rated requirements. Major actions are, for example, the reorganization or formation of a major command (e.g., Air Force Global Strike Command) or a major acquisition program (e.g., growth of the UAS force structure). We recommend requiring that a new appendix on rated requirements be included in each Program Action Directive, the standard planning document for a major action. (See pp. 15–16.)
4. Some of the actions the Air Force can take have primarily long-term effects on the balance between rated inventory and requirements. The recent creation of new career fields for UAS operators and nonrated air liaison officers will eventually alleviate the shortage of rated officers and will provide a substantial increase in the number of people who have sufficient experience to fill rated staff billets, but this will take time. The Air Force could also redesign positions to concentrate tasks that require rated expertise in fewer rated positions and could spin off tasks that do not require rated expertise into new, nonrated positions. (See pp. 16–18.)
5. Projections, especially of requirements, can change rapidly and unpredictably. The aircrew management system must be responsive to avoid having changes throw it out of balance. The system would gain much in responsiveness if it could meet rated requirements while producing and absorbing rated officers at rates below capacity, on average. Maintaining some spare capacity would also help to prevent backlogs of students awaiting training and overmanning of operational units. (See pp. 18–20.)

Logically, spare capacity could be established by either increasing capacity or reducing production and absorption. Options for increasing capacity are generally beyond the scope of this report (but see Taylor, Bigelow, and Ausink (2009) for a discussion of how Air Reserve Components assets—especially highly experienced pilots—could be used to increase absorption capacity). Reducing production may seem unpalatable, as it would reduce future rated inventories, but the process itself will provide ways to cope with those inventory reductions and thus will reduce the problems of doing so. It may be possible that modest reductions in production and absorption could even show a net benefit.

These elements, we feel, could be implemented in the current aircrew management system with little disruption. Various organizations would acquire new responsibilities and/or face changes to some existing responsibilities. But existing responsibilities would not be shifted from one organization to another.

The five actions recommended here do not include an enforcement mechanism, which we feel is needed. It would be advantageous for each owner of rated positions if all owners embraced the process. But individual owners might feel that they could have extra rated officers assigned to them if all other owners embraced the process while they refused to do so. An enforcement mechanism would ensure that all owners live within their rated authorization quotas. (See pp. 21–23.)

Acknowledgments

We express our thanks for the continuing counsel of long-term Air Force aircrew management experts James Robinson (Air Education and Training Command, Deputy Chief Requirements and Resources Division); Craig Vara (Air Mobility Command, Directorate of Operations, Aircrew Operations and Training Division, Chief Force Management Branch); Ed Tucker (Headquarters Air Combat Command, Directorate of Air and Space Operations, Flight Operation Division Chief, Flight Management Branch); Evans Glausier (Headquarters Air Force Special Operations Command, Operation Training Division, Integration Branch); and C. J. Ingram (Headquarters Air Force, Operations, Plans and Requirements, Aircrew Management Branch, Senior Defense Analyst). Many others contributed important information, discussions, and thoughtful reviews, including Col Charles Armentrout (Headquarters Air Combat Command, Directorate of Manpower, Personnel and Services); Col William Watkins (Air Force Personnel Center, Operational Assignments Division); Lt Col Adam Kavlick (Headquarters Air Force, Operations, Plans and Requirements, Director of Operations, Operational Training Division, Deputy Division Chief); Lt Col David "Dewey" DuHadway (Headquarters Air Force, Deputy Chief of Staff Manpower, Personnel and Services, Force Management Policy, Military Force Policy Division); Tom Winslow (Headquarters Air Force, Operations, Plans and Requirements, Aircrew Management Branch, Aircrew Analyst); and Tony Garton (Air Force Personnel Center, Analysis Division).

Thoughtful reviews by Gen (Ret) Paul Hester and Harry Thie helped us improve this report substantially.

Abbreviations

ABCCC	airborne battlefield command and control center
ABM	air battle manager
AETC	Air Education and Training Command
AFB	Air Force base
AFGSC	Air Force Global Strike Command
AFI	Air Force Instruction
AFPC	Air Force Personnel Center
AFSC	Air Force specialty code
ALFA	ALO, FAC (forward air controller), and AETC
ALO	air liaison officer
AMEC	Aircrew Management Executive Council
API	aircrew position indicator
ARC	Air Reserve Components
AWACS	Airborne Warning and Control System
BRAC	Base Realignment and Closure
C2ISREW	command, control, intelligence, surveillance, reconnaissance, electronic warfare
CSAR	combat search and rescue
CSO	combat systems officer
DRU	direct reporting unit
FOA	field operating agency
FTU	formal training unit
FY	fiscal year
FYDP	Future Years Defense Program
HAF	Headquarters Air Force
IFF	Introduction to Fighter Fundamentals
IPT	integrated process team

ISR	intelligence, surveillance, and reconnaissance
JSTARS	Joint Surveillance Target Attack Radar System
MAJCOM	major command
MPA	Military Personnel Appropriation
MPES	Manpower Programming and Execution System
MPESUMD	Manpower Programming and Execution System Unit Manpower Document
PAD	Program Action Directive
PAF	Project AIR FORCE
PAS	Personnel Accounting Symbol
PCS	permanent change of station
PE	program element
PPBE	Planning, Programming, Budgeting, and Execution
PPLAN	Programming Plan
RAQ	rated authorization quota
RSAP	Rated Staff Allocation Plan
SAF	Secretary of the Air Force
SOF	special operations forces
STP	students, transients, personnel holdee
TDY	temporary duty
UAS	unmanned aircraft system

CHAPTER ONE

Introduction

For more than a decade, the Air Force has experienced shortages of rated officers. A decline in force structure (by over 50 percent) since the early 1990s has reduced the capacity of the Air Force to produce, absorb, and develop rated officers. The requirements for rated officers have declined as well, but the reduction in requirements has been proportionately less than the reduction in the capacity to generate new rated officers. Flying billets have declined in proportion to force structure, but nonflying billets have not[1]—the ratio of flying to nonflying billets has gone from 5:1 in fiscal year (FY) 1988 to its current value of 3.8:1.

The number of flying billets is roughly proportional to the number of aircraft in the force structure. To fill all the billets, both flying and nonflying, it has been necessary to increase the number of rated officers per aircraft. Since the late 1990s, when most of the pilots and navigators from the Cold War years had left the service, the Air Force has been producing and absorbing rated officers at the maximum possible rate.[2]

The Air Force makes projections of inventory and funded requirements twice yearly, in April and October. The inventory projection is called the Blue Line and is prepared by the Rated Force Policy Division of the Air Force Directorate of Manpower and Personnel (AF/A1PPR). The projection of funded requirements[3] is called the Red Line and is prepared by the Operational Training Division of Air Force Directorate of Operations, Plans, and Requirements (AF/A3O-AT).[4] The April 2007 projection showed the total requirement for rated officers to be below the total inventory (Figure 1.1), seeming to indicate that the shortage had been eliminated for the foreseeable future.

---

[1] Flying billets are assignments in which flying skills are maintained in the performance of assigned duties. The number of months an officer has accumulated in flying billets in the first 12 and 18 years after becoming rated determines his or her entitlement to continuous Aviation Career Incentive Pay. A nonflying billet neither requires flying nor provides credit towards Aviation Career Incentive Pay.

[2] To suggest that there is a single maximum rate of production and absorption is an oversimplification. The system that produces and absorbs rated officers can be thought of as a network in which each link represents a different step in the several production and absorption processes. Putting a person through a link may require some resources unique to that link and other resources shared by several links, and the availability of those resources will determine the capacities of the links in complicated ways. Different categories of rated officers traverse different paths through the network. Rather than saying production and absorption are maximized, it would be more accurate to say that the throughput of one category of rated officers cannot be increased without sacrificing some throughput of another category.

[3] The Manpower Programming and Execution System (MPES) is the official statement of all military manpower requirements, including rated and nonrated, and active and reserve components. It distinguishes between funded requirements (also called authorizations) and unfunded requirements. The Red Line includes only rated requirements that are funded.

[4] The operational training division is responsible for the day-to-day conduct of Air Force aircrew management matters.

Figure 1.1
April 2007 Projections: No Overall Rated Officer Shortage

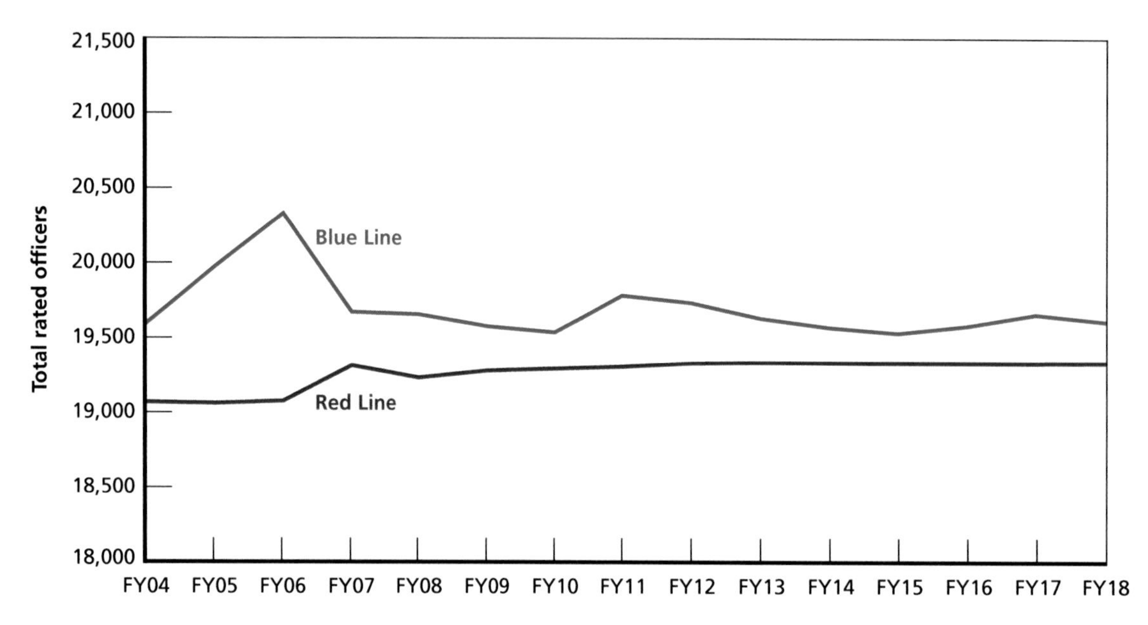

RAND *TR869-1.1*

However, there are many different categories of rated officers with different, though overlapping, skill sets, and no rated officer can fill every requirement. For aircrew management purposes, rated officers are categorized as shown in Table 1.1. A rated officer may be a pilot, a combat systems officer (CSO), or an air battle manager (ABM). The CSO designation incorporates three older training tracks—navigator, electronic warfare officer, and weapon system officer. Pilots and CSOs are further categorized by major weapon system, and ABMs are primarily responsible for command and control.

Separate projections are prepared for each rated category. Even though the FY 2007 Red Line/Blue Line projections showed the overall rated inventory to be in balance with overall requirements, this was not true of each rated category. The shortage of fighter pilots was the

Table 1.1
Categories of Rated Officers as of FY 2009

Pilot	CSO	ABM
Fighter	Fighter	AWACS
Bomber	Bomber	JSTARS
Mobility	Mobility	Ground
C2ISREW	C2ISREW	ABCCC
CSAR	CSAR	
SOF	SOF	
Unmanned	Unmanned	

NOTE: AWACS = Airborne Warning and Control System; JSTARS = Joint Surveillance Target Attack Radar System; C2ISREW = command, control, intelligence, surveillance, reconnaissance, electronic warfare; ABCCC = airborne battlefield command and control center; CSAR = combat search and rescue; SOF = special operations forces.

most serious imbalance (Figure 1.2). The overall inventory and requirements could be in balance only because there were surpluses of mobility pilots.

By the time of the April 2009 projections, the picture had changed drastically (Figure 1.3). In general, one would expect the April 2009 estimates to differ from the April 2007 estimates, because (1) they are based on data for two more years (the estimates for 2008 and 2009 have gone from projections to actual data), and (2) assumptions about future events and conditions will have changed. The 2009 total-inventory curve drops below the 2007 curve between 2008 and 2011 but then converges to essentially the same projections. The fluctuations appear to be the cumulative effect of many small factors and not due to one or two primary factors. Among the reasons identified for the changes in requirements projections were the increase in flying billets resulting from an increase in unmanned aircraft systems (UASs), the MC-12 Liberty buy, new SOF aircraft, and the CSAR replacement project, CSAR-X. Nonflying billets increased due to, e.g., the creation of the Air Force Global Strike Command (AFGSC), AF/A10, and a number of air operations centers.[5]

The picture for fighter pilots also changed (Figure 1.4). In the 2009 projections, requirements dropped in 2010 due to the planned retirement of 208 A-10s, F-15s, and F-16s from the active inventory, a retirement that was not anticipated when the 2007 estimates were made. This drop in requirements closes the gap between requirements and inventory in 2010 and 2011, but it reduces the system's capacity to absorb fighter pilots, so the gap grows in future years as fighter pilots leave the inventory through separation from the active Air Force, promotion to O-6, or grounding.

Figure 1.2
April 2007 Projections: Growing Shortage of Fighter Pilots

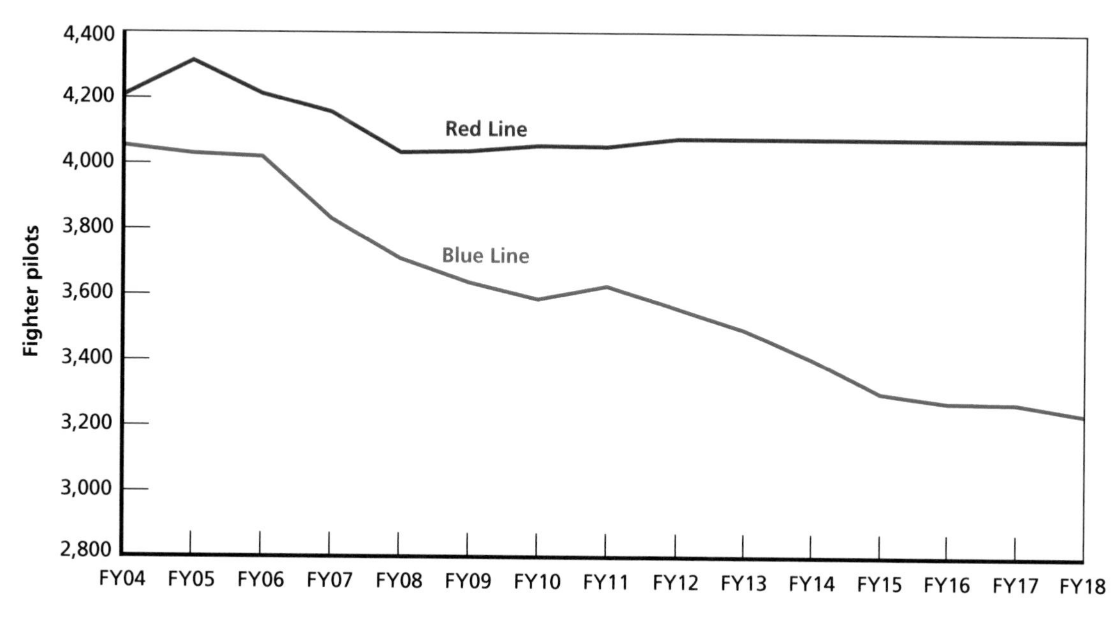

RAND *TR869-1.2*

---

5 The MC-12 Liberty is a manned intelligence, surveillance, and reconnaissance (ISR) platform. New SOF aircraft include the U-28, which is used to support special forces, and nonstandard aircraft intended for airlift. The CSAR-X, now canceled, was a new CSAR helicopter. AFGSC and the Headquarters Air Force (HAF) Strategic Deterrent and Nuclear Integration Staff Office (AF/A10) are both elements of the Air Force plan to reinvigorate the nuclear enterprise.

Figure 1.3
Comparison of Recent Red Lines and Blue Lines

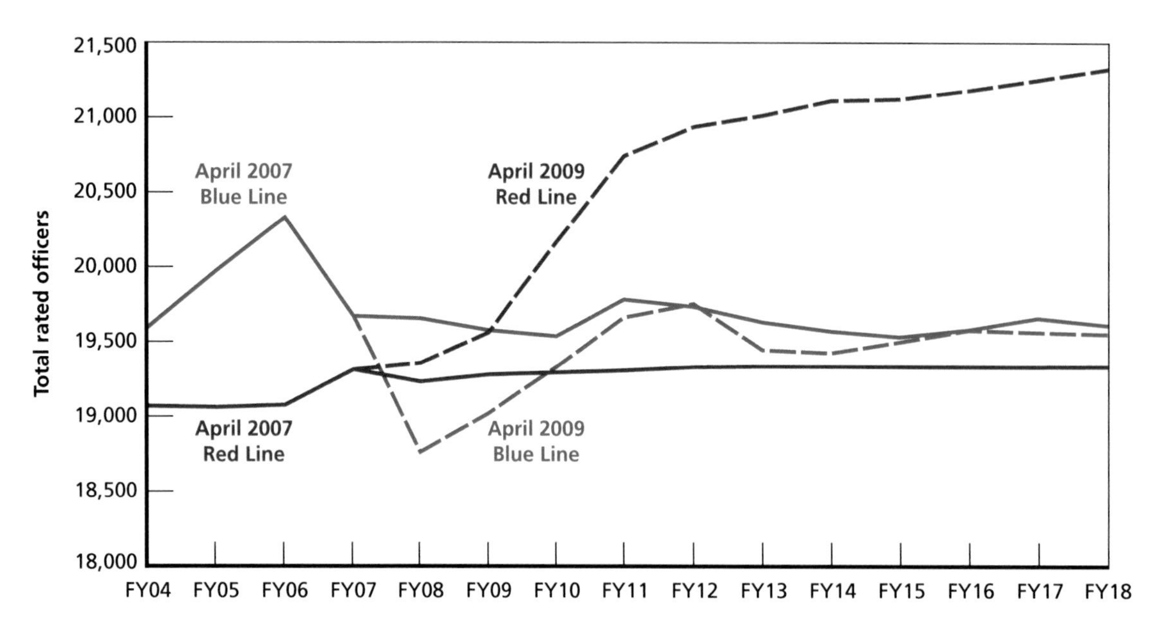

RAND *TR869-1.3*

Figure 1.4
Recent Fighter Pilot Red Lines and Blue Lines

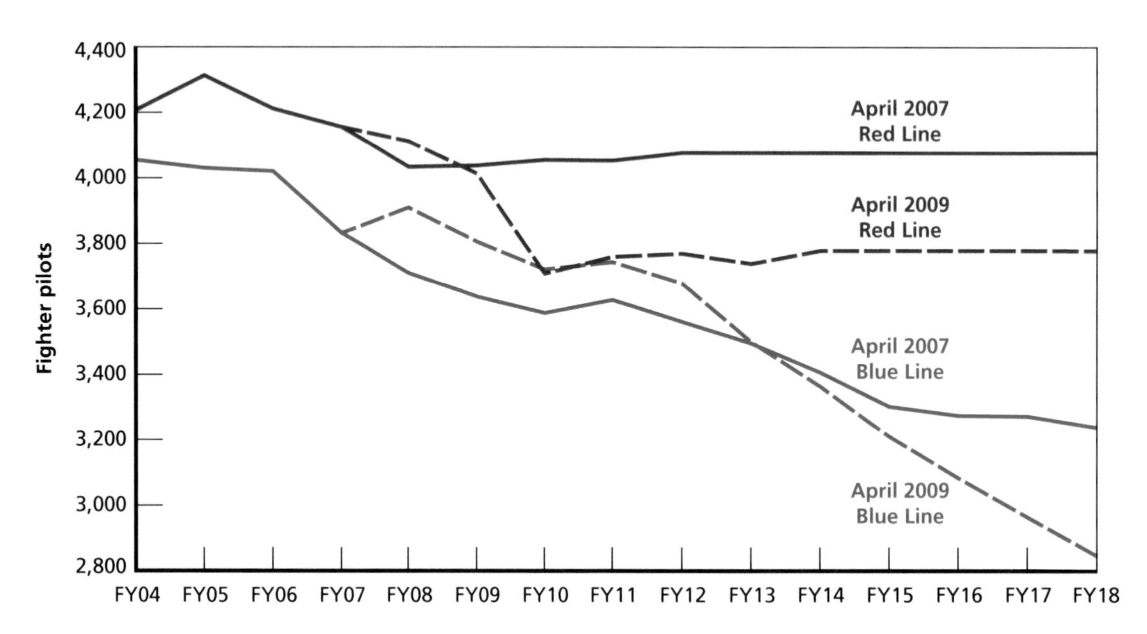

RAND *TR869-1.4*

Impacts of Rated Shortages

At first sight, the overall shortage does not appear large enough to jeopardize the Air Force mission. The projected shortage at the end of FY 2010 is only 4 percent of the total rated requirement, rising to about 8 percent in FY 2013 and beyond. Why does this modest-seeming shortage cause such concern?

The reason is that the shortages are not distributed evenly but are strongly concentrated in certain parts of the Air Force. We have already mentioned that the shortage of fighter pilots is proportionately more extreme than that in other rated categories; the fighter pilot shortage is projected to grow to almost 25 percent of requirements by FY 2018. But the shortage is concentrated in other dimensions as well.

Each rated position in the MPES is classified as a force, training, test, or staff position. Requirements for force, training, and test positions are defined in Air Force Instruction (AFI) 11-412, *Aircrew Management*:

> [Force requirements are] wing-and-below aircrew authorizations for operational flying units. In aggregate these positions comprise the Air Force's aircrew requirements for conducting its operational flying missions. (Paragraph 5.4.3.1)
>
> [Training requirements are] wing-and-below aircrew authorizations for formal flying training units. (Paragraph 5.4.3.2)
>
> [Test requirements are] wing-and-below (or equivalent) aircrew authorizations for test flying units. (Paragraph 5.4.3.3)

Staff requirements are all positions in the MPES that are not force, training, or test positions.[6] In addition to positions in the MPES, the Red Line includes an allowance for students, transients, and personnel holdees (STPs). According to AFI 11-412, paragraph 5.4.4, the STP allowance accounts for

> the average number of aircrew members in advanced student (i.e., aircrew members TDY [temporary duty] to units for education and training course announcement formal flying training), PCS [permanent change of station], professional military education (e.g., IDE [intermediate developmental education] and SDE [senior developmental education]), transient (leave/travel between PCS moves), prisoner, or patient status.

Table 1.2 shows the distribution of requirements among these classes, according to the available data for the end of FY 2008.

The Air Force has no choice but to fill the STP allowance. At any point in time, this number of people will simply not be available to occupy authorized positions.[7] Air Force doctrine directs that 100 percent of force and training positions must be filled, to enable the Air Force to train, experience, and season rated personnel at the maximum possible rate. A significant number of staff and test positions are also considered must-fill billets, leaving the entire shortage to

[6] AFI 11-412, paragraph 5.4.3.4, describes the kinds of positions that are classified as staff/other. As a practical matter, however, every classification scheme must have a place to put things that do not belong elsewhere—an "everything else" class. In this case, the staff classification serves this purpose.

[7] In Table 1.2, STP is an estimate. If the actual number of rated officers in this category exceeds the estimate, fewer people will be available to fill authorized positions. We have no evidence that STP is systematically underestimated, however.

Table 1.2
Rated Requirements by Class as of FY 2008

Class	Number	Percent of Red Line
Force	9,484	49.0
Training	3,514	18.2
Test	702	3.6
Staff	3,827	19.8
Total MPES	17,527	90.5
STP[a]	1,831	9.5
Total Red Line	19,358	100.0

[a] Calculated as the difference between the MPES and Red Line totals.

be distributed among only a few thousand discretionary positions, typically rated staff positions. The overall shortage of 4 percent to 8 percent can become a 40 percent or more shortage in these discretionary billets. When the logic of filling must-fill billets is applied to specific rated categories (especially fighter pilots), even larger shortages occur in other categories.

The shortages are magnified during assignment cycles. There are three assignment cycles per year, and a typical assignment lasts about eight cycles. Naively, then, about one-eighth of the inventory of rated officers should be due for new assignments in any given cycle. But the assignment system will have received requisitions for all of the vacant positions, including the positions just vacated by the officers to be assigned, positions that have been vacant for some time because of the shortage, and any new positions created since the last assignment cycle. All of these factors played a role in the spring 2009 assignment cycle, where only 7 percent (96 out of 1,350) of requisitions for rated staff positions were filled.[8]

In recent assignment cycles, Air Force Personnel Center (AFPC)/DPAO has filled less than 100 percent of requisitions for force and training billets, thus compromising the ability of the Air Force to produce and absorb new rated officers. The willingness to accept smaller future rated inventories in order to fill certain critical staff billets today shows how difficult the trade-offs have become.

The Rated Staff Requirements Integrated Process Team

In February 2009, the Vice Chief of Staff of the Air Force chartered the Rated Staff Requirements Integrated Process Team (IPT) to determine and then recommend courses of action that would balance rated staff requirements with rated inventory and establish an enduring policy and process to maintain this balance. The IPT quickly decided not to simply cut existing requirements, although it did not altogether rule out some cuts. Instead, the IPT assumed that the annual rated requirements review identified valid rated staff requirements—positions that in fact required rated expertise and that were needed to accomplish Air Force missions.[9]

[8] Information provided by the Air Force Personnel Center, Operations Officer Assignments Division (AFPC/DPAO).

[9] This view is not universal. One of our reviewers suggested that changes in the mix of categories of nonflying rated requirements may have lagged changes in Air Force missions and force structure.

However, the IPT did not assume that all of these positions had to be filled by active duty rated officers. Instead, the owners of rated positions[10] were asked to identify positions that could be recategorized to allow them to be filled by other than active rated officers in the category originally specified for each position. Most recategorizations specified that the position would be filled by a civil servant or contractor, with the anticipation that many of these would be retired or separated officers with rated experience. Other options were to fill such positions with nonrated officers, enlisted airmen, or full- or part-time members of the Air National Guard or Air Force Reserve. Or a position could be filled by a rated officer with a different (more abundant) Air Force specialty code (AFSC). In effect, then, the IPT set up an experiment to develop and test a procedure for managing the Red Line.

The first step in this procedure was to tell the owners how many positions they needed to recategorize or, equivalently, how many officers in each rated category they could expect to be assigned to them. These quotas of rated officers were chosen so that requirements and inventory would be balanced by the end of FY 2010. Projections showed that the inventory shortfall should reach a minimum of 834 at that time, so the exercise was less daunting than it would have been if it had addressed larger near-term shortfalls.

In the second step, each owner determined which positions it was willing to recategorize and then met with other owners in a series of four working groups. The first working group comprised owners from HAF, SAF, FOAs, DRUs, and joint agencies. The other three groups, one each for mobility air forces, combat air forces, and SOF, met later. At each working group, owners reported the rated categories for which they could not recategorize enough positions to meet their quotas and the rated categories for which they could recategorize more than enough positions. By matching owners appropriately, it was possible to make trades (e.g., of a fighter pilot for a mobility pilot) that both owners agreed provided a net gain for the Air Force.

After all the groups had met, 836 positions had been recategorized. About 100 too few fighter, bomber, and C2ISREW pilot positions and 69 too few fighter, bomber, and C2ISREW CSO positions were recategorized, but the participants generally agreed that the Air Force had taken a major step toward balancing rated inventory with requirements by the end of FY 2010.

As of this writing, the third step is still ongoing. To make the IPT effort more than a paper exercise, the recategorized positions must actually be filled with the designated types of nonrated personnel. In most of the recategorized positions, a civilian was identified as the replacement for an active rated officer. To enable these positions to be converted, a wedge has been included in the Future Years Defense Program (FYDP) that will fund 352 conversions in FY 2010 and a total of 572 conversions by FY 2013. Each conversion will be accompanied by a compensating reduction in officer end strength.[11] The position owners could take additional rated officer-to-civilian conversions by offsetting them with civilian–to–nonrated officer conversions elsewhere in their organizations. The first civilians were expected to take recategorized positions in November 2009.

For almost all of the remaining positions, an active officer with a nonrated AFSC or a career enlisted aviator was identified as the replacement for a rated officer. The replacements were required to have sufficient experience to hold staff positions (for officers, a rank of major

---

[10] Owners of rated positions include major commands (MAJCOMs), field operating agencies (FOAs), direct reporting units (DRUs), joint agencies, HAF, and the Secretary of the Air Force (SAF).

[11] Personal communication from Thomas Winslow, AF/A3O-AT, February 16, 2010.

or above). An analytical review by AF/A1PF identified a number of AFSCs whose inventories included more staff-eligible personnel than were required of the AFSC.[12]

It was expected that the MPES would be updated to reflect the recategorized positions early in 2010. The summer assignment cycle (June–September 2010) was to fill the first of the recategorized billets with nonrated officers and enlisted personnel.

---

[12] AF/A1PF is the Force Management Division of the Directorate of Force Management Policy. The AFSCs they identified are 13S (space); 15W (weather); 21A (maintenance); 33s (communications); 38F (force support/manpower); 65F (comptroller); 1Ax (career enlisted aviator).

CHAPTER TWO

Elements of an Enduring Process for Maintaining the Balance

A process to maintain the balance between rated inventory and requirements should include the following five actions:

1. Institutionalize a version of the recategorization process pioneered by the IPT.
2. Streamline the processes for converting the recategorized positions. This, too, is a continuation and extension of IPT efforts.
3. Require that the Air Force plan for the effects of major actions, such as the reorganization or formation of a major command (e.g., AFGSC) or a major acquisition program (e.g., growth of UAS force structure), on rated requirements.
4. Consider measures that have primarily long-term effects on the balance between rated inventory and requirements.
5. Develop and maintain some reserve production and absorption capacity. This would increase the flexibility of the aircrew management system for dealing with unanticipated changes in requirements and inventory such as those reflected in Figure 1.3 and, in addition, would help the system operate more smoothly.

Institutionalizing and Regularizing the Recategorization Option

Owners of rated positions (MAJCOMs, FOAs, DRUs, and joint agencies) must conduct an annual review of aircrew requirements.[1] They forward the results of the reviews to AF/A3O-AT, which approves or disapproves them. The reviews ensure that all rated staff positions require rated expertise, but they take no notice of possible inventory shortages. According to AFI 11-412:

> The ability to fill a rated manpower authorization is not a factor in determining whether that rated position is needed. Requirements are established to ensure that the Air Force mission is successfully accomplished.[2] Ensuring appropriate rated expertise is applied to

[1] AFI 38-201, paragraphs 9.2.2.6 and 9.2.3.4.

[2] Public Law 101-189, Sec 633, mandates that "no increase in the number of nonoperational flying duty positions in the Armed Forces (as a percentage of all flying duty positions in the Armed Forces) may be made after September 30, 1992, unless the increase is specifically authorized by law." This mandate is implemented in AFI 38-201, paragraphs 9.2.2.7.3 and 9.2.3.3, which require owners to find "offsetting authorizations for new aircrew staff requirements unless new weapon systems, growth in existing weapon systems, or new aircraft missions generate new requirements." In principle, this could prevent the Air Force from establishing a rated manpower authorization even if such an authorization was needed to ensure that the Air Force mission is successfully accomplished.

the requirement is a separate and distinct function. Documenting rated requirements, particularly during inventory shortage periods, can help make the case for increased resources necessary for increased absorption and proactive inventory/retention improvement initiatives. (Paragraph 5.3.6)

Modification of the Annual Rated Requirements Review

The annual rated requirements review should be modified to take inventory constraints into account.[3] In principle, if the balance between requirements and inventory deteriorated slowly enough, they would need to be rebalanced only once every several years. But the Red Line/Blue Line balance has been volatile in recent years, and the volatility is likely to continue as more legacy fighters are retired, more ISR platforms are purchased, and more remotely piloted vehicles are fielded. We believe that the currently programmed numbers are not the final word.

In advance of the annual review, AF/A3O-AT should provide each owner with an estimate of the number of positions it will be able to fill with rated officers in the coming year. This will be its rated authorization quota (RAQ) and will be distributed over rated categories. Owners of rated positions would still determine that rated expertise of some kind is needed to fill the position effectively. But each owner would have to include a position within its quota to make it an authorized position (i.e., eligible to be filled by an active rated officer). And owners should identify the second-best option for filling each remaining position—civilian, full- or part-time Air Force Reserve or Air National Guard, nonrated active officer, or enlisted person.[4]

The current requirements review also establishes which rated category is required for each position. Owners should consider recategorizing positions from rated categories with severe shortages (e.g., fighter pilot) to categories with greater abundance (e.g., mobility pilot).[5] Recategorizing a fighter pilot position to a civilian position has the same overall effect on the balance between requirements and inventory as recategorizing a fighter pilot position to a mobility pilot position and recategorizing a mobility pilot position to a civilian position. And the latter could have a smaller detrimental effect on performance.

Finally, owners should identify trades they would be willing to make (e.g., a fighter pilot RAQ for a mobility pilot RAQ).

Initially, the current procedure for generating Rated Staff Allocation Plans (RSAPs) could be used for generating the RAQs. Improved methods might be developed in the future. According to AFI 11-412, paragraph 7.5:

---

[3] As mentioned earlier, the MPES distinguishes between funded requirements (also called authorizations) and unfunded requirements. Our suggested modifications to the requirements review are intended to limit funded rated requirements to the number that the Air Force can reasonably expect to fill. They are not intended to affect the number of unfunded requirements at all.

[4] One of our reviewers asked how the rated experience of a contractor or civilian might be kept fresh (we have no suggestions). He pointed out that officers will expect to hold a position for only a few years, while civilians (and enlisted personnel) may expect to occupy the position for a much longer time. The advantage of a longer tenure is that the occupant may learn the job better. The disadvantage is that his or her rated expertise may grow stale.

It is outside the scope of this study to suggest factors that owners ought to consider in determining the second-best option. However, the importance of recent flying experience must vary from one position to another, and this could be such a factor.

[5] Positions could also be recategorized to the generalist pilot category, which calls for any pilot. In today's environment, positions that call for generalist pilots are filled with mobility pilots. Similarly, a position can call for a generalist navigator.

Allocation plans are intended to provide a disciplined, objective approach for the Air Force to bear shortfalls in areas where they can be best mitigated.

The Operational Training Division (AF/A3O-AT) is responsible for producing RSAPs (AFI 11-412, paragraph 1.6.1.1).

The RAQs allocated to an owner should equal the number of rated officers, by category, that owner can reasonably expect to be assigned to it at a specified time, say one to two years in the future. This lead time would give owners advance warning of the number of positions they must find some other way to fill. The IPT effort took place in mid-2009 and aimed to balance requirements with inventory at the end of FY 2010, a lead time of 1.5 years.

Ranking Rated Staff and Test Positions

It would help owners accomplish these new tasks if each owner could rank its staff and test positions (more generally, all positions not considered "must-fill") in the order in which it would prefer to have them filled. All of an owner's positions, regardless of rated category, could be ranked in the same list, including positions that can be filled from more than one rated category (11G and 12G positions, for generalist pilots and navigators, respectively). Position P would be ranked higher than position Q if the owner judged that the value added by filling position P with an active rated officer (rather than with the best available alternative) is larger than the value added for position Q. Ties in rankings would be allowed.

The rankings would be done from the owners' perspectives, not from the perspective of the officers who will occupy the positions. Rated officers will prefer positions that offer exceptional opportunities for professional development and subsequent rapid promotion. But there seems scant reason to expect that owners would tend to rank these positions higher than positions that focus more on their own day-to-day mission. These rankings would not, therefore, establish a new pecking order among rated officers that would affect performance reports and promotion consideration.

Each owner would rank only its own positions. We do not anticipate that the ranked lists of the several owners would ever be combined into a single master list for the entire Air Force. It seems unlikely that the various owners could agree—in finite time—about the importance of filling their own positions relative to those of the other owners. Nor have we detected much appetite within the Air Staff for imposing a solution on the owners. However, given a starting assignment of RAQs to positions—and the current assignment of rated officers to positions amounts to such a starting assignment—the separate lists could be used to improve that assignment, as discussed below.

Given the ranked lists, it would be a simple matter to devise an algorithm that will assign RAQs to positions in such a way that each is assigned to the most highly ranked position possible. All positions without RAQs are candidates for recategorization. Any position without a RAQ that has a higher rank than some position with a RAQ is also a candidate for trade. This can occur if the two positions require rated officers from different categories. For example, the higher-ranked position might require a fighter pilot but be outside the owner's quota of fighter pilots. The lower-ranked position might require a mobility pilot and be inside the owner's quota for mobility pilots. It would benefit the owner to trade an officer in the rated category assigned to the lower-ranked but authorized position for an officer qualified to fill the higher-ranked but unauthorized position.

A clearinghouse would be needed to match owners with complementary candidates for trades. A sensible host for the clearinghouse is A3O-AT, which is responsible for producing the RSAP now and which we nominate to produce the RAQs as well. The clearinghouse will look for trading cycles, groups of owners 1, 2, . . ., n such that owner 1 gives a RAQ for rated category "a" to owner 2, owner 2 gives a RAQ for rated category "b" to owner 3, . . ., and owner n completes the cycle by giving owner 1 a RAQ for rated category "z." For every owner in the cycle to benefit, each owner must have a lower-ranked position to which it has assigned a RAQ for the rated category it is giving up and a higher-ranked position without a RAQ that requires the rated category it is receiving.

The ranked list we have proposed requires only that each owner determine, for any pair of positions P and Q, whether the value added for position P is larger than the value added for position Q. It does not require that the owner estimate the relative sizes of the two values added, i.e., the owner need not determine whether the value added for P is twice as large or only 10 percent larger than the value added for Q. Without knowing relative sizes of values added, the only exchanges that can be assumed with confidence to benefit an owner are exchanges in which the owner receives at least one RAQ for each RAQ given up. Since no owner sees a benefit in giving up more RAQs than it receives, only one-for-one trades are possible.

If, however, owners were able to make defensible estimates of the relative sizes of values added,[6] the lists could be used to identify trades beneficial to the owner where the number of RAQs given up exceeds the number received in exchange. In this case, it would be possible to identify trades of unequal numbers of RAQs that would benefit all participating owners.

Potential Risk

Some Air Force personnel with whom we consulted were concerned that if they recategorized a rated position, they would never be able to reverse the process—they would have permanently lost a rated authorization, and if more active rated inventory became available at some future time, they would be unable to take advantage of it. More indirectly (and perhaps more likely), the capacity to produce, absorb, and season rated officers might increase, perhaps due to the creation of a new rated career field such as UAS operator. Taking rated authorizations off the books might reduce the motivation to use this increased capacity to increase the rated inventory. As a result, the formerly rated position would be filled permanently by someone less qualified than the category of rated person for which it was originally intended.

In the short term, of course, this is no risk at all, for today's choice is to fill the position with somebody who is not active rated or to leave the position empty. But is there a longer-term risk?

We think not. True, there is strong resistance at present to creating new rated requirements, and as long as this resistance remains strong, it would be difficult to restore a rated position to the books. But we attribute the resistance to the fact that there is currently a serious shortage of rated officers. We expect that if this shortage became less severe (the premise of the concern), the resistance to new rated positions would weaken. Indeed, in the history we described in the Introduction, recategorization of rated billets has been a last resort. Logically, restoring rated positions to the books would be the first response to a growth, actual or potential, in the rated inventory.

---

[6] This task would require considerably greater analytic prowess than is needed to merely establish rankings.

Making the Conversion Process More Responsive

Positions that are recategorized must be filled by the designated alternative to an active rated officer in a timely fashion. Otherwise, the positions simply continue to be vacant and the option to recategorize has little value. But actually filling a recategorized position—identifying a qualified person and hiring him or her—is only the last step in the process. We next discuss some of the issues involved in different kinds of conversions.[7]

Conversion to Civil Service or Contractor

As mentioned earlier, the IPT arranged to include funding in the current program objective memorandum for 352 military-to-civilian conversions in FY 2010 and a total of 572 by the end of FY 2013. Some civilians were expected to be on the job by November 1, 2009. Some of the 572 positions are unavailable until after the end of FY 2010, which is when they must be filled in order to avoid having empty rated staff positions.

Moreover, obtaining funding is only one step in a military-to-civilian conversion. A civilian position must be documented as a National Security Personnel System position or a General Schedule position. The documentation must include the duties of the position, the pay to be offered, performance expectations, and required knowledge, skills, and abilities, and it must be completed before recruiting for the position begins. Anecdotes suggest that it can take years to complete the documentation for a position and obtain the necessary approvals. The Air Force Manpower Agency suggests that this time could be compressed by developing standard templates for position documentation and using them as the starting points for military-to-civilian conversions.

The process of converting a rated requirement to a contractor position should be similar to converting one to a civil service position. Administrative details will undoubtedly differ, and different offices in the Air Force and the Office of the Secretary of Defense will oversee the process. Instead of documenting the position as outlined above, one writes a contract. But the contract should by and large include the same elements as the position documentation.[8,9]

Conversion to NonRated Active Military

Converting a rated position to a nonrated position should be the easiest kind of conversion to accomplish. The pay for all active Air Force personnel comes out of the same appropriation category, so there would be no need to arrange a different source of funds. It would be up to an owner to determine which AFSC—the target AFSC—could provide a person qualified to fill the position. The target AFSC could be either an officer or enlisted AFSC.

---

[7] Once a position has been converted, of course, it remains converted indefinitely, assuming it is not reconverted. It no longer appears in the MPES as a rated requirement. Because inventory and requirements projections change from one year to the next, however, we anticipate that additional positions will need to be converted each year for many years to come.

[8] We have in mind only contracting with individuals to perform specific jobs. Contracting with an organization to perform an entire function presents additional issues.

[9] A reverse conversion (contractor or civil servant to rated) should be much easier. Such a conversion would not be made unless there is a rated officer available to fill the position who can do the job better than the civilian. In that case, there will already be money in the military pay appropriation to pay the officer, and he or she will already fit within the statutory end strength. The rest is paperwork and should be no more onerous than that encountered in converting a military position to a civilian one.

However, some analysis will be required to identify which nonrated AFSCs can afford to provide personnel to fill recategorized positions. Only someone with considerable experience (say, 10 years) will be qualified to fill a rated staff position that has been recategorized. Suppose that John Smith, an officer or enlisted person with 10 years of experience in the target AFSC, is assigned to the recategorized position. This will leave a "hole" in the target career field, namely the position that John Smith would otherwise have filled. This hole may be filled by a somewhat less-experienced officer or enlisted person within the target AFSC. The hole that this person leaves will be filled by a third reassignment, and so on until the hole that is left can be filled by an officer just out of initial training in the target AFSC.

Two conditions must be met if this cascade of reassignments is to have only a minor impact on the target AFSC. First, the target AFSC's inventory must exceed requirements in the staff-eligible year groups. Otherwise, each position in the cascade will be filled by someone less experienced (and presumably less qualified) than would have been the case if the rated position had never been recategorized. As mentioned earlier, AF/A1PF has identified a handful of AFSCs for which this is the case.

Second, the target AFSC must have the capacity to train a new member who will fill the last hole in the cascade. Training a new member should require no more than the one- to two-year lead time of the recategorization/conversion process, so the cascade should not provoke a shortfall in the target AFSC.[10] Each career field manager should know how much, if any, reserve production capacity his or her AFSC has.

These factors will also determine how many of its staff-eligible personnel an AFSC can afford to assign outside the AFSC. One limit is its capacity to increase production of new members. But each staff-eligible person assigned outside the AFSC will reduce the average experience in positions that require the AFSC. Too many such assignments will bleed the AFSC of the experienced personnel it needs to fill the positions it owned before it acquired any recategorized rated positions.[11]

Conversion to Air National Guard or Air Force Reserve

There are substantial numbers of highly experienced rated officers in the Air National Guard and the Air Force Reserve—together referred to as the Air Reserve Components (ARC)—so on paper, the ARC should be prime sources of rated expertise.[12] The active forces can tap this resource in various ways: Members of the ARC can, for example, be hired or contracted as civilians, using the same processes as for any other civilian. But because they are in the military, they can also be recalled to active duty, hired temporarily under the Military Personnel Appropriation (MPA) man-day program (AFI 36-2619).

The MPA program offers the active duty Air Force a way to use ARC personnel to assist in performing active duty missions. They are paid from the military personnel appropriation, just as are active duty personnel. Their tours are intended to be short (no more than 139 days annu-

---

[10] If there is no unused capacity, conversion will simply move the vacant position to the target AFSC. This is why stressed AFSCs are poor candidates for conversion.

[11] Of course, if all parties—aircrew managers and the career field manager for the target AFSC—agree that the target career field will lose less than the rated community will gain, they could assign a person outside his or her AFSC anyway.

[12] In principle, if nonrated active personnel (officers or enlisted personnel) qualified to fill some rated positions could be found, qualified nonrated ARC personnel could also be found.

ally, according to AFI 36-2619, but extended to 1,095 days in any four years by Section 416 of the 2005 National Defense Authorization Act [PL 108-375]), but the time limit can be waived.

At the present time, the Air Force does not systematically record the number of people in the MPA program who are in rated positions, nor are the potential future numbers of such people captured in projections of inventory.

Planning for Effects of Major Actions on Rated Requirements

When the Air Force implements a major action, such as standing up the AFGSC, it usually creates new rated requirements. In fact, the growth in rated requirements that occurred between 2007 and 2009 (see Figure 1.3) was primarily due to such actions, including increases in force structure (UAS, MC-12, U-28, nonstandard aircraft) and changes in organization (including AFGSC).[13] While planning the implementation of such actions, the Air Force explicitly considers their effects on total military manpower but does not attempt to separately estimate their effects on rated manpower.

We suggest that the Air Force should explicitly consider effects on rated positions during the planning process. To make this happen, we recommend that an appendix on rated requirements be made standard for the planning documents that help accomplish major actions. These documents are Program Action Directives (PADs) and Programming Plans (PPLANs), defined in AFI 10-501. A PAD is a formal planning document, written at the HAF level, which helps accomplish a major action, such as the reorganization or formation of a MAJCOM, organization, unit, or function, or a new acquisition or modification program. It assigns responsibilities and identifies critical tasks as milestones. A PPLAN also describes Air Force initiatives that help accomplish a major action, but it is written below HAF level; it is usually more specific than a PAD and focuses more on tasks or milestones.

A PAD (we will not mention PPLANs further) consists of a short section describing the background of the plan, its objective, authority or references, assumptions, effective date, and office of primary responsibility. This is followed by as many annexes as necessary, each written by a two-digit office in HAF or SAF, covering the issues for which that office is responsible. For example, Annex A, "Manpower, Personnel, and Services," is written by AF/A1; Annex B, "Intelligence, Surveillance and Reconnaissance," is written by AF/A2; and Annex C, "Operations, Plans and Requirements," is written by AF/A3/5. Because rated management is a shared responsibility of AF/A3O-AT and AF/A1, the new appendix on rated requirements could be an appendix to either Annex A or Annex C. Since AF/A3O-AT would be responsible for managing RAQs, we lean toward Annex C.

The new appendix would provide estimates of the RAQs that new or gaining owners would need in order to implement the major action and would identify the owners that would lose RAQs to offset the gains. Some of the offsets might occur because responsibilities were transferred from one owner to another, but some RAQs might have to be "taken out of hide." The office of primary responsibility would propose these offsets, and AF/A3O would coordinate and adjudicate. The PAD for a major action is not written at the end of the process;

---

[13] Changes in force structure may alter the capacity to produce and absorb rated officers and hence could eventually affect the Blue Line, as discussed in the next section. They have a more immediate effect on rated requirements (the Red Line). Major actions that do not change the force structure generally affect the Red Line but not the Blue Line.

rather, it evolves through multiple versions as the process unfolds.[14] Thus, simply requiring the new appendix would ensure that rated requirements are considered throughout the planning process.

Once the new appendix has been made standard, it may make sense to consider a major action whose purpose is to reorganize the Air Force in order to make more efficient use of rated personnel. Every major command headquarters, every numbered air force headquarters, every air operations center requires a rated presence, so consolidating these organizations would reduce rated requirements. The active and reserve components of the Air Force have separate, duplicate staff functions. Consolidating them should also reduce rated requirements.[15]

Actions with Longer-Term Effects

The elements of the enduring process that we have discussed so far can all influence the balance between rated requirements and inventory fairly quickly (within one to two years). In this section, we discuss actions that will take longer to influence the balance. One such action is the creation of new career fields. The other is redesigning rated positions to make more efficient use of rated personnel.

Creation of a New Career Field

The new UAS operator career field, now undergoing beta test, promises to eventually relieve much of the imbalance between rated requirements and inventory. The new nonrated air liaison officer (ALO) career field will contribute as well. But because the new career fields have no existing inventory, they cannot offer quick relief.

Various estimates have been made of how quickly the Air Force can grow the UAS operator career field, but none is yet official. (The April 2009 Blue Line does not show significant numbers of pilots in the "Unmanned" rated category.) But suppose the Air Force determined that the mature UAS operator career field should be large enough to fill not only the UAS-related force and training positions, but a proportionate share of staff and test positions as well. By our estimates, made in 2008, this would require an eventual inventory of 2,364 officers. Figure 2.1 shows our (optimistic) estimate of how the career field would build to this size.

The top curve in Figure 2.1 is the total. To sustain this total under our assumptions, 169 officers must be produced each year. We assume that an officer, once trained, will follow a career much like that of a fighter pilot:[16] an operational tour, followed by an ALFA (ALO, FAC, and AETC) tour, then a second operational tour, after which he or she would be qualified to occupy a staff position.

Currently, most of the UAS operator positions are filled by rated officers on their ALFA tours. By 2015, a UAS operator career field evolving as in Figure 2.1 could liberate about 500

---

[14] We have Draft version 14 of the PAD for AFGSC.

[15] Further exploration of this issue is beyond the scope of this project but could be undertaken in future research.

[16] The UAS operator we assume here has an active duty service commitment of six years rather than the 10-year commitment of a fighter pilot. We adjusted the cumulative continuation rates so that our UAS operator has a total active rated service of 14 years, essentially the same as that of a fighter pilot.

Figure 2.1
Notional Evolution of the UAS Operator Career Field

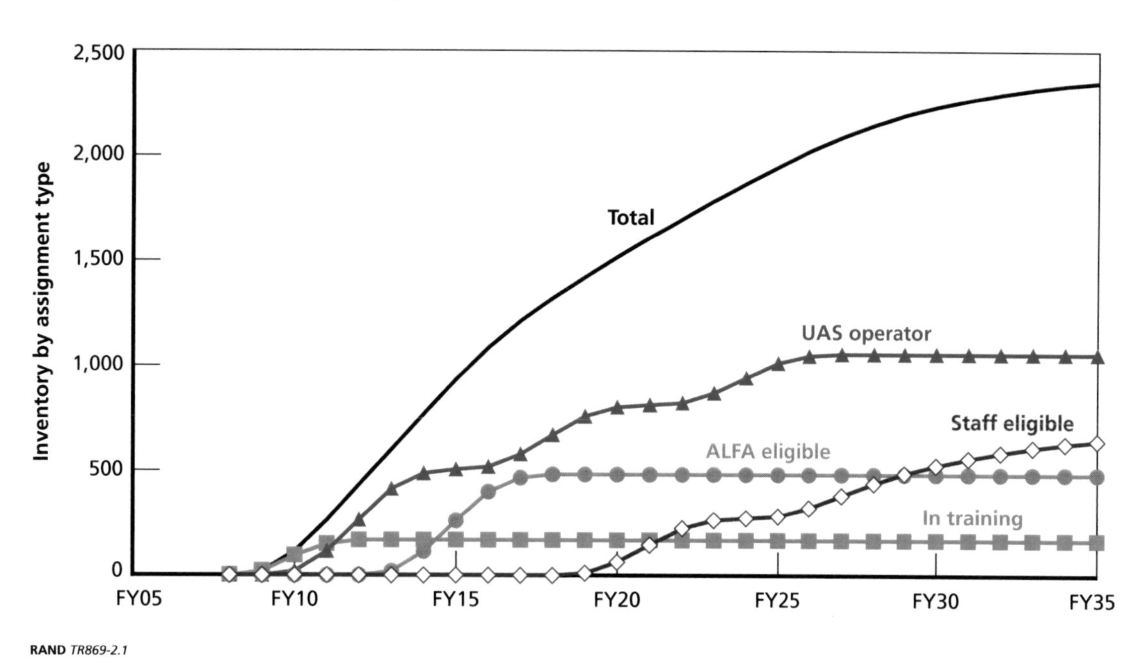

RAND *TR869-2.1*

of those officers for assignments in their own career fields. But no career UAS operators could be qualified for staff positions until 2019.[17]

Redesign of Rated Positions

Next we consider the possibility of redesigning rated positions so that the tasks that require rated expertise are concentrated in fewer positions, and tasks that do not require rated expertise are spun off to new, nonrated positions. Positions could also be redesigned to concentrate tasks requiring the kinds of rated expertise peculiar to rated categories with severe shortages. In the first case, the total rated requirement would decrease. In the second, the total rated requirement might remain the same, but the requirement for a particularly stressed rated category would decrease.

We speculate that fruitful opportunities for redesigning positions might occur for different reasons in offices that have very few positions than in offices that have many positions. The September 2008 version of the MPES contained 3,827 rated staff positions. As shown in Figure 2.2, 304 of those positions were in Personnel Accounting Symbol (PAS) codes that had only a single rated staff position. At the other end of the spectrum, 2,204 positions were in 74 PAS codes that had more than 10 rated staff positions each. Here we use PAS code as a surrogate for an owner and a location. That is, people assigned to positions in the same PAS code should be working in the same place and for the same organization.

[17] Which rated staff positions will these 10-year UAS operators be qualified to fill? Their presence will improve the balance between requirements and inventory far more if they are qualified, or can be made qualified, to fill positions coded for the rated categories with the most extreme shortages (e.g., fighter pilots).

Figure 2.2
Size Distribution of Offices with Rated Staff Positions

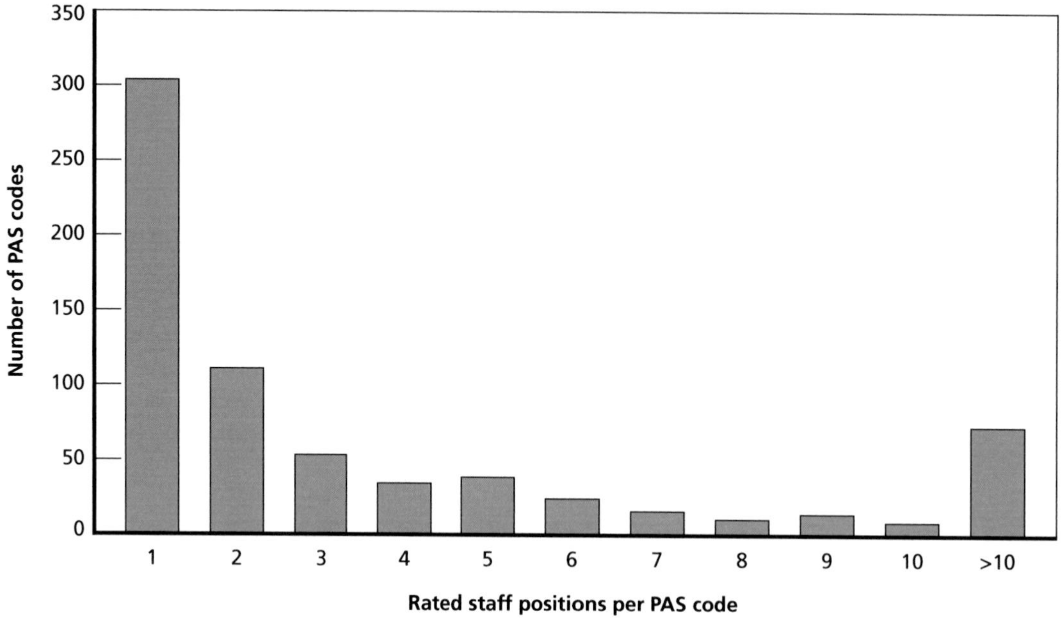

SOURCE: MPES, as of September 2008.
RAND *TR869-2.2*

First, consider PAS codes with a single rated staff position. If there is to be any rated presence at all in a PAS code, one rated person is the smallest number that can be assigned. This is true even if the tasks requiring rated expertise occupy only a modest fraction of that person's time and effort. If a way could be found to provide that rated expertise remotely, it might be possible to consolidate those positions into a substantially smaller number. Undoubtedly, some of these positions require a physical rated presence, but some savings might be possible.

Next, consider PAS codes with many rated staff positions. People in many of these offices work side by side. They can ask one another questions and work together on projects. In some of these offices, the rated officers should be able to specialize in providing rated expertise to their nonrated colleagues. This could reduce the number of rated positions needed in the office.

Developing and Maintaining Some Spare Production and Absorption Capacity

As noted earlier, since the late 1990s, the Air Force has been producing and absorbing rated officers at the maximum possible rate. If this could be changed—if the aircrew management system could produce and absorb rated officers at rates below capacity—the Air Force could operate more smoothly and respond more flexibly to unanticipated changes in requirements and inventory such as those reflected in Figure 1.3. Reducing production and absorption rates[18]

[18] In principle, this could be done by investing in greater production and absorption capacity, but addressing this possibility is beyond the scope of the present report.

of the various categories of rated officers would lower the inventory projections, and this would in turn necessitate converting more requirements from rated to nonrated.

Benefits of Spare Capacity

Suppose the aircrew management system had been operating at less than full production and absorption capacity when the new requirements for experienced rated officers shown in Figure 1.3 emerged. A new rated officer could be produced and assigned to a tour in an operational squadron in only one year. Then a rated officer with the necessary amount of experience could be moved from that squadron to the rated staff position. The net effect is that the operational force would become slightly less experienced. In effect, the operational billets would provide a reservoir of experienced rated officers that could be replaced by less-experienced officers (including newly produced officers) and assigned to staff work.

Operating without spare capacity has important costs. To use every available space in a training class, aircrew managers must ensure that every time a space becomes available, a student is on hand to fill it. Historically, however, factors that aircrew managers can neither control nor predict affect the numbers and timing of both arriving students and spaces available for them in the various training classes. Queueing theory tells us that as these uncertainties grow larger, system behavior deteriorates and backlogs accumulate at ever-lower values of the utilization factors (e.g., see Hillier and Lieberman, 2001).

This happened in FY 2008, in large part because capacity in the Introduction to Fighter Fundamentals (IFF) course was lost during the move—mandated by the 2005 Base Realignment and Closure (BRAC)—from Moody Air Force Base to Randolph Air Force Base. By September 2008, the backlog had grown to 99 fighter pilots, who awaited space in an IFF course for an average of 252 days.[19] Because there was (and is) no spare IFF capacity, the production of fighter pilots had to be reduced for FY 2009 and FY 2010, so the system could work off the backlog.[20]

A new pilot is absorbed during his or her first operational tour. An operational unit does not have a hard and fast capacity to absorb new pilots. Rather, every pilot who graduates from the formal training unit (FTU) can be immediately assigned to an operational unit. If the FTU graduates too many pilots from the basic course, the operational units become simultaneously overmanned and underexperienced. An extreme example of this occurred in FY 2000 in the A-10 squadrons at Pope AFB (Taylor et al., 2002). Because overmanning had made it necessary to share the limited number of sorties among too many pilots, 60 percent of the squadrons' primary mission pilots had been decertified from combat-mission-ready status.

Developing Spare Capacity

Logically, to establish spare capacity to produce and absorb rated officers, one must either increase capacity or reduce production and absorption. While it is beyond the scope of this report to consider how capacity might be increased, we have discussed one possible option, i.e.,

---

[19] Figures are from a briefing, "Pilot Break in Training (BIT)," presented by James "Robbie" Robinson, Deputy Chief, Requirements and Resources Division, AETC, at AMEC 09-1, October 22–23, 2008.

[20] The pilots in the backlog were considered to be rated officers and therefore were included in the Blue Line. Because they had not completed their training, however, they were not qualified to fill any requirements. In effect, they had to be "assigned" to STP.

using ARC assets (especially highly experienced pilots) to relieve absorption pressures in the active component (Taylor, Bigelow, and Ausink, 2009).

If production of rated officers is decreased, there will eventually be a drop in the inventory. In the reverse of a process described above, there will also be an immediate reduction in experienced officers available to fill staff positions, since there will be fewer new, inexperienced officers available to fill line positions in operational units (which are must-fill positions), and their places will be taken by more-experienced officers. As expected, then, reducing production exacerbates the shortage of experienced rated officers, and for that reason the senior Air Force leadership has resisted this measure in the past (see Taylor, Bigelow, and Ausink, 2009, pp. 10–12).

In the past, however, the Air Force had no established process for reducing rated requirements in response to inventory shortfalls. Thus the entire cost of reducing production would be felt as an increase in vacant rated staff positions. With the implementation of an enduring process, the costs can be experienced in other, less painful ways, while the benefits of reduced production (discussed above) will be as large as ever. Perhaps resistance to reducing production might lessen. Then again, it might not.

Monitoring Spare Capacity

To maintain spare capacity, it is necessary to measure and manage capacity in relation to workload. Adequate information to do this is already collected routinely, although it is not used explicitly for this purpose.

Utilization factors (the ratio of workload to capacity) of the various training courses could be used to track spare production capacity. The Air Education and Training Command (AETC) routinely estimates undergraduate pilot training and FTU capacity. In typical queueing models, the average size of the backlog is inversely proportional to 1 minus the utilization factor and hence increases without bound as the utilization factor approaches 1. Research would be needed to identify critical values for the utilization factors that would minimize the cost of backlogs plus the cost of unused capacity.

Manning levels, experience mix, and flying hours per inexperienced aircrew member in operational units would serve to track spare absorption capacity. Data on all of these quantities are already collected and reported for the purpose of measuring and managing operational readiness.

To maintain a comfortable reserve of absorption capacity (in fighter squadrons, at least), previous work (Taylor, Bigelow, and Ausink, 2009) suggests keeping unit manning below 105 percent of authorized and the experience level (ratio of experienced aircrew position indicator [API]-1 pilots to total API-1 pilots) at 60 percent or above. The experience target we recommend is substantially higher than the target the Air Force has historically used for fighter squadrons (50 percent or even lower).

CHAPTER THREE

Responsibilities and Enforcement

Responsibilities

We see the responsibilities for carrying out the various elements of the enduring process described in this report distributed as follows: AF/A3O will publish RAQs annually, as it now publishes the RSAP, and will distribute them to owners of rated positions.[1] It will also adjust the RAQs as necessary for trades the owners are willing to make. As necessary, it will write the new appendix to Annex C of each PAD that may levy new rated requirements.

The owners of rated positions—including MAJCOMs, FOAs, DRUs, joint agencies, HAF, and SAF—will annually review the rated positions they own, ensuring that the funded positions do not exceed the RAQs they are given in any rated category. They will also identify the positions they wish to recategorize and the trades of quotas they are willing to make.

The AF/A1M organizations at HAF and at MAJCOMs will implement conversions of recategorized positions.

The assignment process will be unchanged. However, the enduring process will ensure that shortages are smaller, so the assignment process will become less contentious.

The Aircrew Management Executive Council (AMEC) or a subgroup of it will monitor the aircrew management system. It will use the indicators suggested earlier and/or others to assess whether the system is in balance. If it is not, the AMEC or subgroup will recommend corrective courses of action.

Figure 3.1 shows how the enduring process could be implemented with minimal disruption of the current management system. Various organizations would acquire a few new duties and/or change a few existing duties, but existing responsibilities would not be shifted from one organization to another.

The Enforcement Mechanism

The hammer in Figure 3.1 represents the enforcement mechanism. Each owner of rated positions would be better off, we think, if all the owners embraced this process. But an owner might feel that it would be better off still—i.e., that it could arrange to have extra rated officers assigned to it—if all other owners embraced the process and it refused to do so. The hammer would prevent this from happening.

[1] We propose annual publication of RAQs, rather than publication on some other schedule, simply because the current rated requirements review occurs annually. If an annual process is tried and found wanting, the schedule can be changed.

Figure 3.1
An Enduring Process for Reconciling Rated Officer Inventory and Requirements

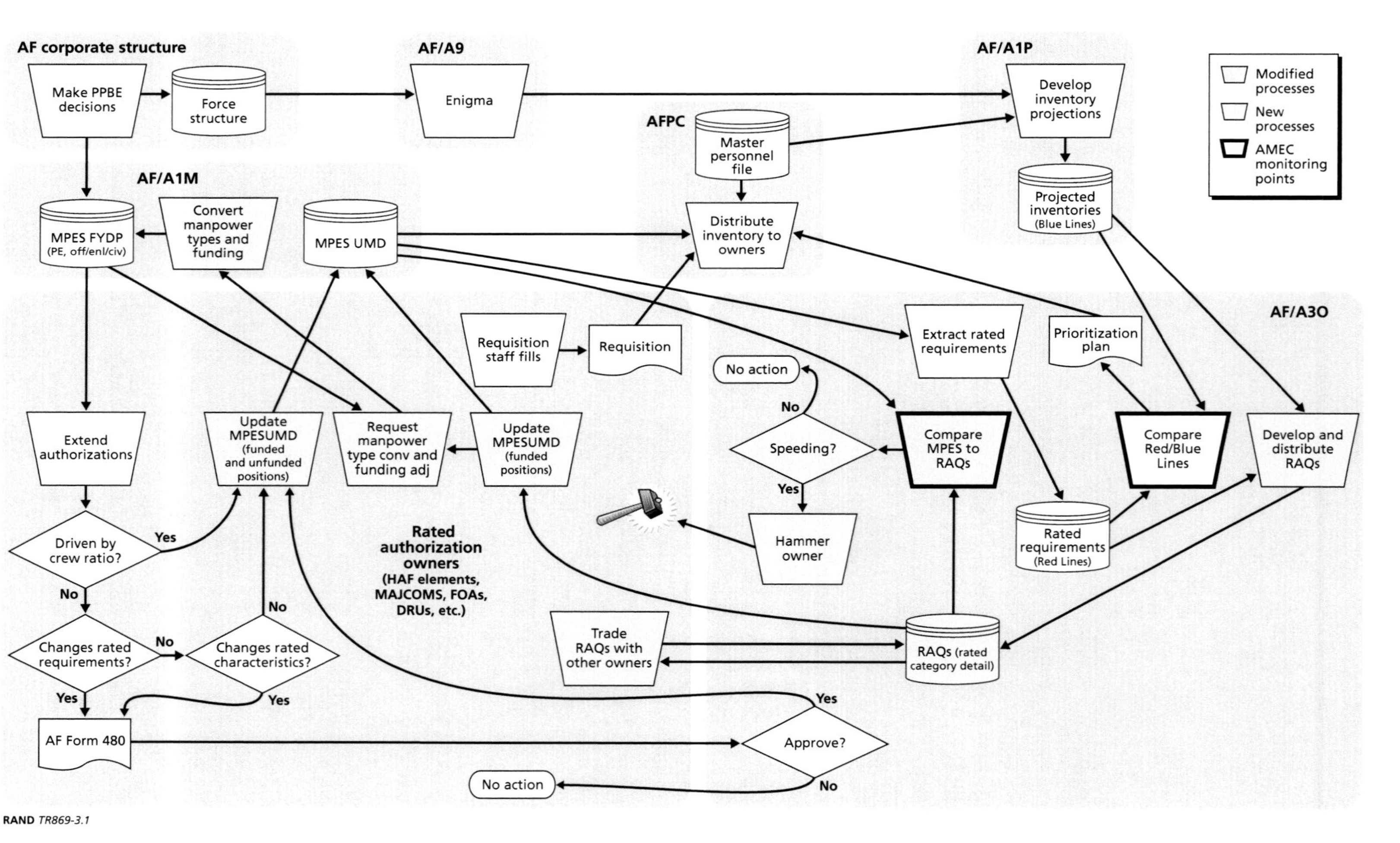

We have identified two points in the enduring process at which enforcement mechanisms might operate. The first is during assignment, which is when the current process most visibly breaks down. Currently, the Directorate of Operations (AF/A3O) provides the Directorate of Assignment at the Personnel Center (AFPC/DPA) with the RSAP, against which AFPC/DPA assigns the available rated inventory. The RSAP establishes fill rates for various categories of rated positions, but (unlike our proposed RAQs) not owner by owner. During each assignment cycle, owners attempt to persuade AFPC/DPA to grant exceptions to the RSAP priorities. We have seen no documentation of the degree of success they achieve, but all the owners we spoke to consider this to be a significant problem.

Under the enduring process, the RAQs will play the role of the RSAP. AFPC/DPA will know each owner's RAQs and can use them as constraints on the number of rated officers assigned to each. Consistently imposing such constraints will require support from senior leadership, but it should be easier to resist making exceptions than it is under the RSAP. The fact that RAQs are specified owner by owner makes it possible to identify the owner who pays for an exception.

Moreover, owners should have less cause to seek exceptions during the assignment process, because they will have had time and opportunity during the annual rated requirements review to lobby for changes in their RAQs. We have described how owners could improve their RAQ holdings through trading. They could also make a RAQ holding more palatable by converting rated positions. Owners should therefore be more satisfied with an assignment that adheres to the RAQs than one that adheres to the current RSAP.

The second point at which an enforcement mechanism might operate is the annual requirements review. The success of the enduring process depends on each owner ensuring that its authorized rated positions in each rated category do not exceed its RAQs. If an owner fails to meet this requirement by a certain date, AF/A3O and/or AF/A1M could apply a default procedure to "deauthorize" the necessary number of that owner's positions. The support of senior leaders would be needed to make such deauthorizations stick.

Bibliography

Air Force Instruction 10-501, *Program Action Directives (PAD) and Programming Plans (PPLAN)*, Department of the Air Force, January 5, 1994.

Air Force Instruction 11-412, *Aircrew Management*, Department of the Air Force, December 10, 2009

Air Force Instruction 36-2619, *Military Personnel Appropriation (MPA) Man-Day Program*, Department of the Air Force, July 22, 1994.

Air Force Instruction 38-201, *Determining Manpower Requirements*, Department of the Air Force, December 30, 2003.

Hillier, Frederick S., and Gerard J. Lieberman, *Introduction to Operations Research*, New York: McGraw-Hill, 2001.

Public Law 101-189, National Defense Authorization Act for Fiscal Years 1990 and 1991.

Public Law 108-375, Ronald W. Reagan National Defense Authorization Act for Fiscal Year 2005.

Taylor, William W., James H. Bigelow, and John A. Ausink, *Fighter Drawdown Dynamics: Effects on Aircrew Inventories*, Santa Monica, Calif.: RAND Corporation, MG-855-AF, 2009. As of October 10, 2010: http://www.rand.org/pubs/monographs/MG855/

Taylor, William W., James H. Bigelow, S. Craig Moore, Leslie Wickman, Brent Thomas, and Richard S. Marken, *Absorbing Air Force Fighter Pilots: Parameters, Problems, and Policy Options*, Santa Monica, Calif.: RAND Corporation, MR-1550-AF, 2002. As of October 1, 2010: http://www.rand.org/pubs/monograph_reports/MR1550/